BEGINNERS GUIDE TO ELECTROCUTTURE FOR PROFIT

From Voltage to Harvest: Navigating the Electrocutture Landscape for Agricultural Excellence

Lucille Sarron

Table Of Contents

CHAPTER 1

Electrocutture

In agriculture, electrocutture is a new paradigm that uses electrical and technical developments to improve many facets of farming operations. This novel strategy maximizes production, resource efficiency, and sustainability by fusing cutting-edge electrical applications with conventional agricultural practices. Comprehending the origins, development, and significance of Electrocutture offers a significant understanding of its capacity to transform contemporary agriculture.

Definition And History: The application of electrical and electronic technology to agricultural operations with the goal of improving production, accuracy, and efficiency is known as electrocutture. The idea came about when it was realized that electricity might be used to solve a number of problems that traditional agricultural methods were facing. With the development of technology came the realization that electrical principles might be used in agriculture, which gave rise to the field of electrocutture.

Evolution Of Electrocutture: The history of electrocutting dates back to the invention of the first electrical devices and their use in farming. The first stage was characterized by the introduction of electrically driven equipment including irrigation systems and tractors. Sensor, automation, and data analytics advancements over time helped Electrocutture develop into a more complex and networked system.

Precision agriculture, which uses tools like GPS, drones, and smart sensors to accurately manage resources like water, fertilizer, and pesticides, has emerged as a major component of electrocutture in recent years. Better yields, less of an impact on the environment, and higher efficiency are the results of this progress.

Importance In Contemporary Agriculture: Because electrocuture has the ability to solve significant issues facing the sector, it is extremely important in contemporary agriculture. Among the crucial elements are:

1. Precision farming is made possible by electrocuture, which allows farmers to carefully monitor and regulate elements including soil moisture, temperature, and nutrient levels. This focused strategy reduces waste and maximizes resource use.

2. Automation and robots: Using electrocutture to combine automation and robots in agriculture increases productivity for operations like planting, harvesting, and pest management. This lowers the need for labor and raises overall productivity.

3. Data-Driven Decision-Making: To give farmers useful insights, electrocutture uses data analytics. Farmers may optimize yields and avoid losses by using data analysis to make educated decisions about weather patterns, soil conditions, and crop health.

4. Sustainability: Electrocutture helps to promote sustainable agricultural practices by using resources more effectively and having a smaller negative impact on the environment. This is in line with the increasing emphasis on ecologically friendly farming practices throughout the world.

To sum up, electrocutture is a revolutionary method of farming that uses electrical technology to improve productivity, accuracy, and sustainability. As the area develops further, it seems to be crucial in solving the problems facing contemporary agriculture and guaranteeing food security for an expanding world population.

CHAPTER 2

The Fundamentals Of Electrocutting

It appears that there may be some confusion over the term "Electrocutture." "Electroculture" often refers to a contentious and untested theory that says low-level electrical currents may be used to improve agricultural yields and plant development. It's crucial to remember, nevertheless, that the majority of mainstream scientific research does not corroborate the claims made for electroculture.

Considering that you intended "Electroculture," let's examine the fundamentals of this idea:

The Electroculture Fundamentals:

1. Definition: The process of applying low-level electrical currents to plants in order to promote growth, increase production, and maybe strengthen their resistance to pests and diseases is known as electroculture.

2. Historical Context: Although the idea of electroculture has been around for a while, scientists are still debating it. While some supporters assert beneficial impacts on plant development, others contend that the data for electroculture is conflicting and inadequate.

Electroculture System Components:

1. Electrodes: In most electroculture systems, electrodes are either affixed to the plants or buried in the ground. The electrical currents that are supplied to the plants are delivered by these electrodes.

2. Power Source: In order to produce the electrical currents, a power source is required, such as solar panels or a low-voltage generator. In order to guarantee that the plants receive the desired low-level stimulation, the power supply needs to be carefully controlled.

3. Control System: To regulate the strength and duration of the electrical currents, an electroculture system needs a control mechanism. Without harming the plants, this management system helps to maximize the growing environment for them.

4. Monitoring Tools: Plant growth rate, crop production, and general health are some of the characteristics that may be used to gauge how successful electroculture is.

Electricity's Role In Plant Growth:

1. Stimulation of Metabolism: Electroculture proponents assert that low-voltage electrical currents can activate the plant's metabolic pathways, resulting in a rise in the absorption of nutrients and general development.

2. Enhanced Nutrient Absorption: Electroculture has the potential to increase plant health and robustness by improving the absorption of nutrients from the soil.

3. Resistance to Pests and Diseases: According to some supporters, electroculture can strengthen a plant's innate defenses against pests and illnesses. It is believed that defensive systems activate in reaction to the electrical stimulation, causing this to happen.

4. More Research is Needed: Although anecdotal evidence points to the advantages of electroculture, more thorough and controlled investigations are required to provide a comprehensive grasp of its effectiveness and its uses, according to mainstream scientific literature.

In summary, electroculture in agriculture is still a speculative field that requires further study to support its promises and prospective advantages. Until definitive data is obtained, it is important to investigate more conventional methods of maximizing plant development and to approach this idea with a critical perspective.

CHAPTER 3

Organizing Your Electrocutting Configuration

An inventive kind of farming called electrocutture incorporates electricity and technology into customary agricultural techniques. To guarantee a successful and effective operation, careful consideration of a number of elements must be taken into account while planning your electrocutture arrangement. These are important points to remember:

Evaluating Your Area

Examine the area you have available for your setup before beginning any electrocutture. Think about the following:

1. Size and Layout: Take into consideration the available space while designing the layout of your electrocutture equipment. This covers the layout of

the crops, the electrical fixtures, and the maintenance routes.

2. Sunlight Exposure: Evaluate how much light is coming into each section of your space. Certain crops require optimal sunshine to flourish, thus knowing this can help you choose and arrange your crops.

3. Access to Resources: Since irrigation systems are frequently a crucial part of electrocutture, make sure that water supplies are easily accessible. If necessary, evaluate the availability of additional resources, such as shade.

Selecting The Right Crops

A key factor in the success of your electrocutture endeavor is crop selection. When selecting crops, keep the following in mind:

1. Climate Suitability: Select plants that will flourish in the climate where you live. While certain electrocutture settings may involve controlled

surroundings, planning will benefit from the knowledge of the surrounding environment.

2. Growth Requirements: The amounts of water, light, and nutrients that different crops require vary. Make sure the crops you select can live in harmony with one another in your electrocutture system.

3. Market Demand: Take into account the amount of demand there is for the crops you want to grow. To guarantee a successful endeavor, match your crop selection to customer tastes and market trends.

Comprehending The Needs For Electricity

Electricity is essential to electrocutture in order to control and maximize crop development. Recognize the electrical needs of your setup:

1. Select the kind of lighting system that will work best for your crops. For instance, LED grow lights

use less energy and emit certain light spectrums that are good for plant development.

2. Climate Control: Put in place measures to regulate the humidity and temperature, particularly inside. HVAC (heating, ventilation, and air conditioning) systems may be involved in this.

3. Automation: Look at technology that can automate processes like monitoring, fertilizer supply, and irrigation. Automation aids in resource optimization in addition to increasing efficiency.

Setting Aside Money For Electrocuture

A realistic budget must be created in order to execute electrocutture successfully. Think about the following elements:

1. Infrastructure Costs: Take into consideration the expenses related to establishing the physical infrastructure, such as the buildings, the electrical wiring, and the support systems.

2. Equipment and technologies: Set aside money for necessary items like sensors, grow lights, automation systems, and any particular electrocuture technologies.

3. Operational Costs: Project your continuous expenses for energy, upkeep, and any consumables needed for your electrocutture system.

4. Contingency Planning: Set aside money in your budget for unforeseen costs or last-minute adjustments to your original plans. Being adaptable is essential for dealing with unanticipated obstacles.

The basis for a viable and prosperous electrocutture endeavor is laid by thoroughly organizing your setup with an emphasis on space evaluation, crop selection, electricity requirements, and budgeting. Over time, consistent monitoring and tweaks will further optimize your system.

CHAPTER 4

Choosing Appropriate Equipment

Applying electrical currents to plants to improve growth, yield, and general plant health is known as electroculture, a cutting-edge and contemporary method of farming. Selecting the appropriate equipment is essential to the effective operation of electroculture systems. This tutorial examines important factors to take into account when choosing equipment, including different kinds of electroculture systems, lighting choices, devices for controlling humidity and temperature, and automation tools for effective administration.

1. Types Of Electrocutture Systems:

The applications and designs of electrocutture systems differ. Farmers may select the best system for their unique requirements by being aware of the many varieties. Typical electroculture setups consist of:

• Direct Current (DC) Systems: These systems rely on an electrical current that flows continuously. Plant growth in general is known to be enhanced by DC systems, which also improve nutrient absorption and root development.

• Systems with Alternating Current (AC): In AC systems, the direction of the electrical current is periodically reversed. These systems are prized for their capacity to boost blooming, enhance pest and disease resistance, and accelerate plant metabolism.

• Pulsed Electromagnetic Field (PEMF) Systems: These systems increase photosynthesis, improve nutrient absorption, and improve the general health and vigor of plants. They do this by using pulsed electromagnetic fields.

The selection of the electroculture system that best suits the cultivation's goals will be determined by assessing the unique requirements of the crops as well as the intended results.

2. Lighting Options: Plant development depends on proper lighting, and electroculture frequently entails maximizing light exposure. Think about the following choices for lighting:

• LED Grow Lights: These lights are a popular option for electroculture since they are energy-efficient and adjustable. They enable producers to modify light spectra to suit the particular needs of various plants at various stages of growth.

• Fluorescent Lighting: This type of lighting is affordable and appropriate for small-scale electroculture systems. They offer an excellent range for plant growth, especially for young and seedling plants.

• High-Intensity Discharge (HID) Lighting: High-intensity illumination is provided by HID lighting, which includes metal halide and high-pressure sodium lamps. Larger electroculture enterprises can use them.

To maximize the results of electroculture, the light spectrum and intensity must be matched to the particular requirements of the plants.

3. Devices For Controlling Temperature And Humidity:

Precise management of environmental conditions is typically advantageous for electroculture systems. Humidity and temperature are crucial for the growth of plants. Take into account the following tools for efficient control:

• Climate Control Systems: These systems manage the growth environment's humidity and temperature. In order to provide the ideal environment for plant development, they could contain humidifiers, air conditioners, and heaters.

• Ventilation Systems: To avoid heat accumulation and preserve ideal conditions, there must be sufficient airflow. Systems for ventilation maintain a fresh supply of CO_2 while assisting in temperature and humidity regulation.

• Smart Sensors and Controllers: By integrating sensors and controllers, temperature and humidity parameters can be monitored and adjusted in real-time, providing a stable and favorable environment for electroculture.

4. Tools For Automation For Effective Management:

Automation is essential to improving electroculture operations. The following resources support effective management:

• Automatic Irrigation Systems: These systems minimize water waste by ensuring that plants receive the appropriate amount of water at the appropriate time, thus fostering optimal development.

• Smart Monitoring Systems: These systems enable producers to measure many factors in real-time, including plant health, nutrient levels, and soil moisture, by utilizing sensors and linked devices.

• Robotic Cultivation Tools: By utilizing sensors and actuators, autonomous robots may carry out manual labor-intensive operations like harvesting, trimming, and weeding, hence increasing overall efficiency.

Using automation technologies can save labor costs, increase resource utilization accuracy, and simplify electroculture procedures.

To sum up, choosing the appropriate tools is essential to the effective use of electroculture. Growers may manage their environment to stimulate plant development, enhance yields, and optimize the benefits of electroculture technology by carefully selecting the type of electroculture system, lighting options, temperature and humidity control devices, and automation tools.

CHAPTER 5

Configuring Your Device For Electrocutions

Using regulated electrical stimulation to cultivate plants is a novel technique called electrocutture, which presents a potential way to maximize yields and improve farming methods. We will go over installation tips, wiring concerns, safety precautions, and space optimization for optimal yield in this comprehensive guide to electrocutture system setup.

1. Installation Instructions:

Choosing the Proper Site: Make sure your electrocutture system is situated in a readily accessible and well-ventilated area. Make sure you are close to a dependable power supply and take humidity and temperature into account. For plants to flourish, there must also be enough illumination, either artificial or natural.

Installing Electrode Arrays: For electrical stimulation to be successful, electrode arrays must be installed correctly. For plant type and development stage-specific array installation, adhere to manufacturer specifications. To encourage consistent plant reactions, make sure the electrodes are distributed evenly.

Connecting Control Units: In accordance with the system requirements, carefully connect control units to the electrode arrays. Verify the wiring again to make sure there are no loose connections that might affect how well the electrocutture system works.

2. Electrical And Safety Procedures:

Electrical Wiring: Make sure the wiring in your electrocutture system is insulated and of the right grade for its electrical needs. Use a color-coding scheme to make wire identification simple, and make sure all connections are properly grounded to avoid electrical risks.

Surge Protection: To protect your Electrocutture system from power fluctuations and surges, use surge protectors. In order to safeguard electronic components and guarantee the reliability of the electrical stimulation process, this is essential.

Emergency Shutdown Procedures: Make emergency shutdown switches easily visible and establish emergency shutdown procedures. Learn these protocols so that you and your team can react quickly in the event of an emergency.

3. Maximizing Yield By Optimizing Space:

Plant Layout and Spacing: Arrange your Electrocutture system to make the most use of available space. Take into account the distinct needs of every kind of plant, such as placement and alignment, in order to optimize exposure to electrical stimulation and improve total productivity.

Vertical Farming Techniques: To maximize space, learn about vertical farming techniques. Plant density may be raised and yield per square foot can be enhanced by making use of vertical space.

Environmental Controls: Establish the perfect growing environment by using environmental controls like humidity and temperature regulation. In the end, these controls improve plant health and increase the electrocutture system's efficacy.

In conclusion, careful planning for installation, wiring, safety precautions, and space optimization is necessary while setting up an electrocutture system. Following these recommendations will help you cultivate a more favorable environment for plant development, which will eventually result in increased yields and the effective incorporation of electrocutture into contemporary agriculture.

CHAPTER 6

Recognizing Plant Nutrients For Electrocutting

Using state-of-the-art technology, electrocutture is a novel approach to agriculture that has completely transformed conventional agricultural practices. Understanding and improving plant nutrition to increase crop development and production is a crucial component of electrocutture. In this situation, customized formulations, monitoring strategies, and nutritional solutions are essential to the effectiveness of electrocutture.

Formulations & Solutions For Nutrients:

1. Hydroponic Systems: Hydroponic systems, in which plants grow in nutrient-rich water solutions rather than soil, are frequently used in electrocuture. Creating the right nutrient solutions is crucial to

giving plants the vital components they require. This entails being aware of the unique needs of every crop and modifying fertilizer amounts accordingly.

2. Macro and Micronutrients: Both macro and micronutrients must be included in nutrient solutions. While micronutrients like iron, zinc, and copper are essential in lesser levels, macronutrients like nitrogen, phosphorous, and potassium are needed in higher proportions. Reaching the ideal equilibrium is essential for the health and growth of plants.

3. PH Control: The nutrition solution's pH level has a big impact on the nutrients' availability. Plants can effectively absorb nutrients if the pH range is kept within an ideal range. Automated pH control mechanisms are commonly included in electrocutture systems in order to regulate the nutrition solution and avoid any imbalances.

Keeping An Eye On Nutrient Levels:

1. Sensor Technologies: To precisely track nutrition levels, electrocutture uses cutting-edge sensor technologies. Nutrient concentration in the solution is measured with the use of nutrient sensors, electrical conductivity (EC) meters, and other monitoring equipment. With real-time data, dietary excesses or deficiencies may be avoided right away.

2. Data Analytics: Farmers may examine patterns in nutrient data over time by integrating data analytics into Electrocutture equipment. With this data, nutrient formulations may be optimized, crop-specific requirements can be determined, and future plant cycles can be planned with knowledge.

Modifying Nutrient Solutions According To Type Of Crop:

1. Crop-Specific Requirements: At different stages of growth, different crops have different nutritional needs. Comprehending these variances is vital in customizing nutritional solutions. Optimizing resource use can be achieved by programming electrocutture devices to modify nutrient concentrations according to the unique requirements of each crop.

2. Seasonal Modifications: Based on changes in the weather, electrocutture enables dynamic modifications to nutrient solutions. Nutrient requirements vary with changes in the environment. In order to make sure that plants get the proper nutrients at the right times, automated systems can adjust to these changes.

In summary:

A thorough understanding of nutrient solutions, accurate monitoring, and crop-specific adaptations are necessary to comprehend plant nutrition in electrocuture. In a world that is changing quickly, Electrocutture helps to promote high-yield, sustainable agriculture by utilizing cutting-edge technology to improve fertilizer delivery efficiency. The future of contemporary agriculture will be shaped by further study and innovation that will deepen our understanding of plant nutrition in electrocutture.

CHAPTER 7

Controlling Environmental Elements

It appears that there may be a miscommunication in your request. The name "electrocutture" seems to be a combination of "electro" and "agriculture," implying a relationship between agriculture and electricity. But electricity isn't usually used in traditional agricultural methods.

Please elaborate or define the word if you're referring to a particular field or technology that combines these components.

Assuming you are interested in the overall idea of managing environmental conditions in agriculture, the following topics will be covered: managing pests and diseases, ventilation systems, and temperature and humidity control:

1. Managing Humidity And Temperature:

• Significance: The development of plants is greatly influenced by temperature and humidity. Every kind of plant has a range of ideal temperatures and humidity levels for growth.

• Techniques: To manage humidity and temperature, greenhouses or controlled-environment agriculture (CEA) systems can be used. To produce the perfect atmosphere for plant development, these systems may include technology for controlling humidity, temperature, and heating.

• Technology Integration: Real-time environmental condition monitoring and adjustment is possible through sensors and automated systems, guaranteeing that crops are kept at the ideal temperature and humidity levels.

2. Techniques For Ventilation:

• Significance: Ensuring enough ventilation is crucial for preserving air quality, averting the accumulation of pathogens, and facilitating photosynthesis via gas exchange.

• Natural Ventilation: Especially in greenhouses, utilizing natural airflow through open doors and vents may assist in keeping plants in a healthy environment.

• Mechanical Ventilation: To maintain a steady and regulated airflow in larger agricultural settings, mechanical ventilation systems could be required. Fans, louvers, or other mechanical devices could be involved in this.

3. Managing Insect And Disease Problems In Agriculture:

• Integrated Pest Management (IPM): Using an IPM strategy entails integrating chemical, physical, cultural, and biological control techniques to manage pests in an eco-friendly manner.

• Biological Controls: Instead of using chemicals to manage pest populations, natural predators or parasites that prey on certain pests can be introduced.

• Crop diversity and rotation: Altering crop varieties and planting schedules can sabotage insect life cycles and lower the likelihood of disease accumulation in the soil.

4. Integration Of Technology:

• Precision Agriculture: Farmers may monitor their fields more efficiently by using technology like sensors, drones, and satellite imaging. This makes it possible to use resources precisely, minimizing waste and negative environmental effects.

• Data-driven Decision Making: Farmers may optimize their agricultural methods for higher yields and sustainability by using data on temperature, humidity, insect populations, and disease prevalence to influence their decisions.

Whether they choose to practice traditional farming, greenhouse cultivation, or a more technologically advanced method like electrocutture—if that term refers to a particular field or technology in agriculture that involves electricity—farmers can establish a more controlled and sustainable agricultural system by carefully managing these environmental factors.

CHAPTER 8

Crop Upkeep And Care

Crop care and maintenance are essential for maintaining the health and production of plants in the field of electrocutture, a cutting-edge agricultural technique that combines modern technology with conventional farming methods. This idea covers a number of important topics, such as training and pruning plants, keeping an eye on growth patterns, and identifying stress and deficiency signs. To comprehend each of these elements' importance in relation to electrocutture, let's take a closer look at each one.

1. Plant Pruning And Training: In order to maximize plant efficiency, electrocutture requires the use of pruning and training techniques. Sophisticated robotic systems with sensors and instruments for precise work are used for focused pruning, which eliminates unwanted branches, leaves, or fruits. This improves the crop's overall look while also encouraging improved airflow and solar penetration, which lowers the risk of disease and maximizes photosynthesis.

Furthermore, automatic training systems are used to drive plant development in the required directions. Electrocutture uses sophisticated algorithms and sensors to track the growth of plants, modifying the training procedure to maximize productivity and conserve resources. This creative method guarantees that plants are positioned for effective harvesting and that every plant gets the materials it needs to flourish to its full potential.

2. Tracking Crop Growth Patterns: To measure crop growth patterns in real-time, electrocutture uses state-of-the-art monitoring devices. Sensors that are buried in the ground, on plant surfaces, and in the surrounding environment gather information on a variety of characteristics, including temperature, nutrient content, and moisture content. Artificial intelligence systems process this abundance of data to produce insights into the growth and health of crops.

Farmers in Electrocutture can make data-driven decisions to modify irrigation schedules, fertilizer treatments, and other cultivation procedures by regularly monitoring growth trends. Improved yields and resource efficiency result from this proactive strategy, which guarantees that crops receive the best care possible throughout their growth cycle.

3. Identifying Stress And Deficiency

Signs: One of electrocutture's advantages is its capacity to identify and treat stress and nutritional shortages early on. Sophisticated sensors and imaging technologies are used to examine minute variations in the size, color, and structure of plants.

Electrocutture enables focused treatments, including modifying nutrient levels, putting preventive measures in place, or executing precise irrigation, by quickly diagnosing these problems. Lowering the need for excessive pesticide or fertilizer usage and minimizing crop losses, promotes ecologically friendly and sustainable farming methods.

In conclusion, Electrocutture's method of crop care and maintenance is at the forefront of innovation in the rapidly changing agricultural world. Farmers can now prune, train, monitor, and identify stress and deficiency symptoms with unprecedented accuracy because of the combination of smart technology, automation, and data analytics. This all-encompassing method contributes to the sustainability and effectiveness of contemporary agricultural techniques while guaranteeing the well-being and yield of crops.

CHAPTER 9

Procedures For Harvesting And Post-Harvesting

Using electrical stimulation to boost plant growth and productivity, electroculture is a relatively novel and sustainable way of farming that incorporates elements of conventional farming practices. Practices related to harvesting and post-harvest handling are essential to optimizing the advantages of electrocutture. In the framework of electrocutture, let's explore the essential ideas of calculating harvest periods, harvesting methods for various crops, and post-harvest management and storage.

1. How To Calculate Harvest Times For Electrocutting:

• Electrocutture is the process of applying regulated electrical stimulation to plants in order to affect their physiological functions. Plant electrical reactions, such as variations in conductivity or bioelectrical signals, are monitored to determine harvest periods in electrocutture. Sensors can be used by farmers and researchers to assess plant factors that are altered by electrical stimulation, which can assist in determining the best times to harvest.

2. Methods Of Harvesting For Various Crops In Electrocutture:

• Recognizing that various crops may react differently to electrical stimulation is the field of electrocutture. Harvesting methods must be customized to meet the unique needs of every crop. As an illustration:

• Leafy Greens: In electrocutture, electrical stimulation may improve the nutritious content of leafy greens like spinach or lettuce. Mature leaves can be harvested by carefully selecting them, leaving the plant whole to support future development.

• Fruits: Electrical signals can be utilized to enhance the size and quality of fruits on fruit-bearing plants. In order to establish a fruit's maturity, harvesting procedures may entail evaluating its color, size, and hardness.

3. Handling And Storing After Harvest In Electrocutture:

• In electrocutture, post-harvest procedures are essential to maintaining the nutritional content and quality of crops.

• Electrical stimulation and shelf life: By controlling metabolic processes, electrical stimulation can affect a crop's post-harvest shelf life. It is possible to slow down degradation and preserve freshness during storage by using controlled electrical conditions.

• Temperature and Humidity Control: Electrical systems for monitoring and managing temperature and humidity in storage facilities may be included in electrocutture farms. This keeps harvested vegetables fresher longer by halting fungus development and degradation.

• Electrostatic Technologies: For post-harvest processing, certain electrocutture systems may make use of electrostatic technologies. Harvested crops can be made cleaner and less contaminated by microbes by using electrostatic charges.

In conclusion, by integrating electrical stimuli into the growing, harvesting, and post-harvest processes, electrocutture adds a distinctive dimension to conventional agriculture. Optimizing harvesting periods, using crop-specific harvesting methods, and putting improved post-harvest handling and storage procedures in place all help to maximize the advantages of electrocutture and guarantee effective and sustainable farming operations in the future.

CHAPTER 10

Promoting The Electrocutture Products You Make

To stand out in the market and draw in your target audience in the fast-paced world of electrocuture, good marketing is essential. In order to effectively market your electrocutture goods, this book will go over important ideas including determining your target market, developing branding and packaging strategies, and using persuasive selling tactics.

1. Choosing The Market You Want To Enter:

The first stage in developing an effective marketing plan for your electrocutture products is identifying your target market. Think about doing these actions:

a. Market Research: To determine potential clients' tastes and purchasing patterns, carry out in-depth market research. To obtain insights, examine competition strategy and industry trends.

b. Using psychographics, behavioral patterns, and demographic data, create thorough client profiles. This will assist in focusing your marketing efforts on particular customer groups.

c. Determine specialized markets inside the larger electrocutture sector by using niche targeting. Marketing campaigns can be made more individualized and successful by concentrating on particular market segments.

2. Packaging And Branding:

In a crowded market, building a solid brand presence is critical. Good packaging and branding will set your electrocutture items apart from the competition:

a. Brand Identity: Create a distinctive brand identity that embodies the principles and characteristics of your electrocuture goods. This involves coming up with a catchy logo, selecting eye-catching hues, and developing an engaging brand narrative.

b. Sustainable Packaging: To reflect the principles of today's consumers, highlight environmentally friendly packaging. Stressing sustainable practices improves your company image and draws in environmentally sensitive customers.

c. Visual Appeal: Make an investment in eye-catching packaging to showcase the excellence and creativity of your electrocutture goods. Perceptions of consumers may be greatly influenced by visually striking designs.

3. Marketing Techniques For Products Using Electrocutting:

After determining your target market and building a solid brand, use these powerful selling techniques to increase product sales:

a. Online Presence: Create a powerful online presence by utilizing e-commerce platforms. Make use of social media, e-commerce platforms, and your own website to expand your audience and make your products easily accessible.

b. Teach customers about the advantages and special qualities of your electrocutture goods through educational marketing. To establish credibility and trust, provide educational information on social media, blogs, and videos.

c. Partnerships & Collaborations: Take a look at partnering with merchants, influencers, or brands that complement each other. Forming alliances might help you reach a wider audience and market your electrocutture products.

d. Limited-Time Offers and Promotions: To generate a sense of urgency and drive quick transactions, use time-limited specials, discounts, or promotions.

To market your electrocutture goods effectively, you must take a calculated strategy to determine your target market, build a solid brand, and put clever selling techniques into practice. You may position your electrocutture products for success in a competitive market by knowing your customers, developing a strong brand, and using a variety of sales channels.

CHAPTER 11

Troubleshooting Typical Problems

The creative use of electrical stimulation in agriculture, or electroculture, has demonstrated encouraging outcomes in terms of increasing crop growth and production overall. Electroculture systems, like any technical system, can, nevertheless, run into a number of problems that need to be troubleshooted. In this talk, we'll look at three important facets of dealing with difficulties in electroculture: identifying and resolving electrical issues, tackling issues unique to certain crops, and handling system malfunctions.

1. Identification And Resolution Of Electrical Issues:

Ensuring the correct operation of electrical components is one of the essential requirements for sustaining a successful electroculture system.

Common electrical problems might be caused by bad wiring, malfunctioning power supplies, or interruptions in the supply of electrical impulses. To solve these issues:

a. Perform Regular checks: You can find any problems before they get worse by conducting routine checks of power sources, wiring, and electrical connections. A steady electroculture system may be ensured and electrical disturbances can be avoided with routine maintenance.

b. Use Diagnostic instruments: Using multimeters and other diagnostic instruments can help identify the source of electrical issues. It is essential to quickly identify and replace malfunctioning parts in order to keep the electroculture setup functioning properly.

c. Put Redundancy Measures in Place: To lessen the effect of individual component failures, incorporate redundancy measures into the system architecture. This might entail redundant wiring, fail-safe systems, and backup power sources.

2. Taking Care Of Crop-Specific Issues:

Crops might react differently to electrical stimulation, thus it's important to know each crop's specific needs in order to maximize the results of electroculture. To solve issues unique to crops:

a. Investigate Crop Electro-Sensitivity: Find out how sensitive a certain crop is to electricity. By knowing how various crops react to electrical stimulation, electroculture techniques may be adjusted to optimize advantages for each kind of plant.

b. Modify Electrical factors: Depending on the needs of the crops being grown, adjust electrical factors like voltage and frequency. Ensuring that the electroculture system is optimal for the intended plants requires periodic monitoring and modifications.

c. Think About Crop Rotation tactics: To avoid cumulative stress on certain crops, implement crop rotation tactics. This can lessen the negative effects of extended electrical stimulation on some plant species and preserve the health of the soil.

3. Handling Failures In The System:

Systems for electroculture may have unanticipated failures in spite of preventative efforts. Creating plans to deal with system malfunctions is essential to reducing downtime and possible crop loss. In the event of a system failure:

a. Create Comprehensive Emergency Protocols: To direct actions in the event of a system breakdown, create extensive emergency protocols. Clearly outline the procedures for separating impacted parts, evaluating damage, and starting fixes or replacements.

b. Invest in Monitoring Systems: Implement modern monitoring systems that can offer real-time data on the state of the electroculture setup. This makes it possible to identify any problems early and take proactive measures to address them before a failure happens.

c. Training and Readiness: Make certain that staff members engaged in electroculture activities get adequate training in troubleshooting techniques. Training may help you become more prepared, which can cut down on how long it takes to find and fix system errors.

Maintaining a profitable and long-lasting farming operation requires solving typical electroculture problems. using the resolution of electrical issues, comprehension of crop-specific difficulties, and application of contingency plans for system malfunctions, professionals may maximize the development of crops and boost agricultural output using electroculture systems. A thorough approach to troubleshooting, proactive maintenance, and routine monitoring are essential components for maximizing the potential of electroculture in contemporary agriculture.

Summary

In summary, the exploration of the field of electrocutture has been both thrilling and life-changing. It is clear as we get to the end of our investigation into this cutting-edge subject that combines technology and agriculture that electrocutture has the potential to completely change the way we farm. With the promise of greater production, sustainability, and efficiency, the use of electronics, sensors, and automation in conventional agricultural operations has opened up new possibilities.

Thinking Back On Your Experience With Electrocuture:

Think back for a moment on your experience with electrocuturing and the advancements in utilizing technology to improve farming methods. Whether you are an experienced farmer or a computer fanatic, you have probably learned a lot, experimented a lot, and had to adjust along the way. Through the use of automated gear and the installation of smart sensors, Electrocutture has given farmers the ability to make data-driven choices, maximize resource efficiency, and reduce their environmental effects.

Future Directions For Electrocutting:

The field of electrocuture has much more promise for the future. Expect more developments in the fields of machine learning, artificial intelligence, and precision agriculture. Predictive analytics applications, robotics for precise planting and harvesting, and drone integration for aerial monitoring will all become more widespread. The development of electrocutture will be crucial in tackling issues like food security, climate change, and the rising need for environmentally friendly farming methods.

Concluding Remarks For Successful Electrocutture Farming:

Take into account the following guidance to succeed in the field of electrocuture and guarantee a successful endeavor:

1. Invest in Education: Remain up to date on the most recent advancements in farming technology. To improve your knowledge and abilities, go to webinars, seminars, and workshops.

2. Customize Solutions: Adapt electrocutting techniques to your unique agricultural requirements. Since every farm is different, using a tailored strategy guarantees the best outcomes.

3. The key is data management; embrace decision-making based on data. Organize and evaluate data from sensors and other electronic equipment in an

effective manner to learn about crop health, soil quality, and resource use.

4. Work Together and Build Your Network: Assist other practitioners, researchers, and business leaders in the field of electrocuture. Networking and cooperation may result in beneficial alliances, information exchanges, and access to leading-edge technology.

5. Sustainable Practices: Include sustainable agricultural methods in your strategy for Electrocutting. Think about how your activities affect the environment and work to strike a balance between embracing technology and being environmentally conscious.

6. Remain Flexible: The area of electrocuture is ever-evolving, with new innovations appearing on a regular basis. Remain flexible and prepared to adjust to the constantly changing agricultural technology scene.

In summary, electrocutture is more than simply a farming technique; it's a dynamic, ever-evolving ecosystem that necessitates a fusion of conventional knowledge with cutting-edge invention. Farmers that embrace this fusion may grow crops as well as a wealthy and sustainable future for agriculture.

THE END

www.ingramcontent.com/pod-product-compliance
Lightning Source LLC
Chambersburg PA
CBHW070042260726
48658CB00002B/697